ISBN 978-1-5272-2945-7

For inquiries scan QR code

Hang on a minute **Silly Human**, we need trees.
Let's find out why. And what global warming is all about.

Trees breathe in CO_2 and breathe out oxygen.
We don't want too much CO_2 in our atmosphere.

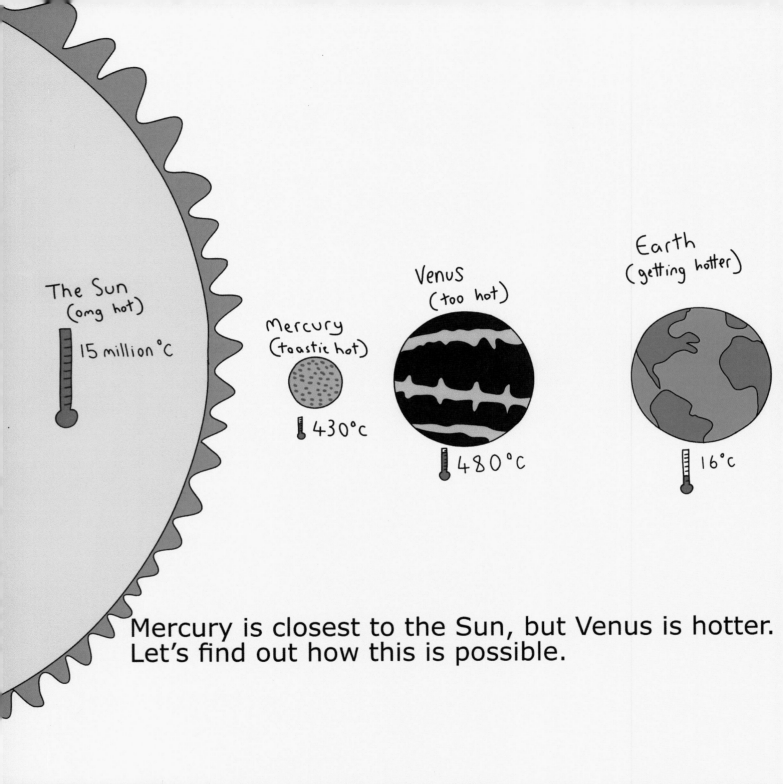

Mercury is closest to the Sun, but Venus is hotter.
Let's find out how this is possible.

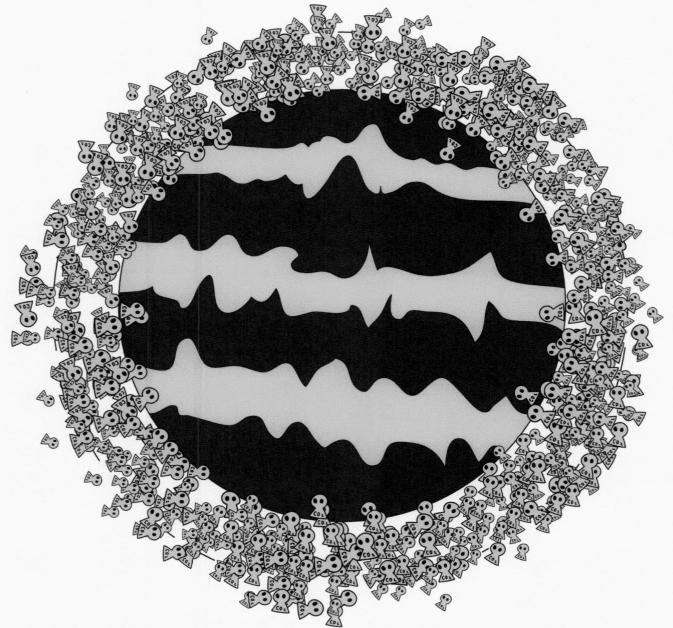

An atmosphere is a layer of gas around a planet.
Venus has an atmosphere of 96% CO_2.
CO_2 traps heat, so nobody can live there.

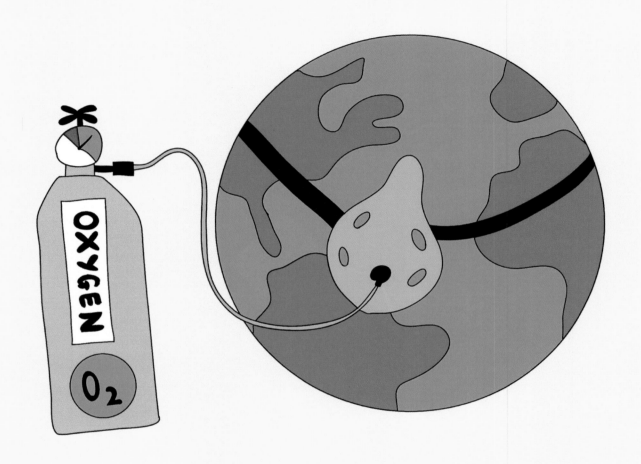

More CO$_2$ means less oxygen on Earth.
Let's find out how **Silly Human** is making CO$_2$.

Our planet was working perfectly until **Silly Human** came along. (who is basically a chimpanzee wearing pants).

He thought it would be a good idea to cut down trees.

Then he burns them, burning things creates more CO_2.

He also burns fossil fuels. These are millions of years worth of old plants and dinosaurs squashed underground.

He makes them into petrol then puts it in his car.
And what does a car engine do? Burns it... **Silly Human**.

He turns fossil fuels into petrol, in a factory.
And what does the factory do? Burns it... **Silly Human**.

When he puts his phone on charge or turns the lights on, this uses electricity. Which is made by... Burning things.

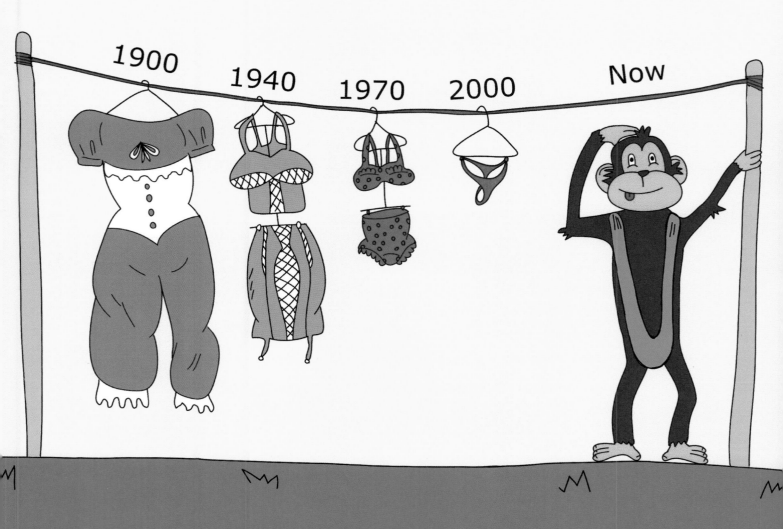

So all this is adding more CO_2 to our atmosphere.
Which means our planet is getting hotter.
Here's some proof.

As the Earth gets hotter, the ice in the North Pole is melting.
If it all melts, where are polar bears going to live?

But don't worry, we still have time to change all this.
Let's see what **Silly Human** can do to help our planet.

Planting trees is a good start. When they grow big, they will munch lots of CO_2.

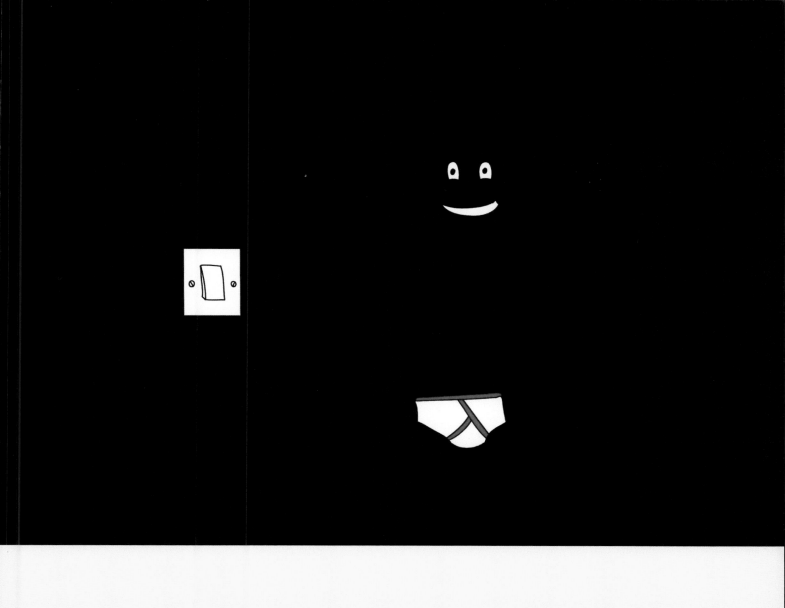

Try to use less electricity, turn the lights off when you
don't need them.
I said when you don't need them... **Silly Human**.

Reading a book is good, no electricity needed.
Although a chair would be useful.

Here are some ways to make electricity, without making CO_2.

Solar panels.
They use the Sun's energy to power your devices.

Wind farms.
When the wind spins them, they make electricity.
Not that sort of farm... **Silly Human**.

Try to use the car less, walk, ride your bike or go to the park.
I bet you would be good at the monkey bars.

Printed in Great Britain
by Amazon